Sensory Enhancements for a More Connected World

Designing for Safety - Ensuring Protection in Quad-Party Open AI Creations

Life Winks LLC

The Future of Victim Proof Quad-Party Interfaces: What Lies Ahead for Vulnerable Populations?

Contents

Blurb VIII

Dedication X

Preface XI

Introduction XIII

1. Data as a Tool for Empowerment: Creating Inclusive Interfaces for Vulnerable Individuals 1

2. Two-Party Interfaces: The Challenges of Creating Safe and Inclusive Digital Spaces 25

3. Battered Concrete Fences: Overcoming Systemic Barriers to Inclusive Interfaces 52

4. Self-Reflective Detection Data for Social Justice: The Ethics of Creating Two-Party Interfaces 81

5. Preventative Protocols: Designing for Safety Proofing Two-Party Open AI Co-Creations 113

6. The Future of Inclusive Interfaces: What Lies Ahead and Within for Vulnerable Populations? 144
7. Enriching Open Data Circumstances for Vulnerable Individuals 174
Afterword 211
Acknowledgments 212
About Author 215
Also By 218

Blurb

"This book explores the intersection of technology and disability, offering insights and strategies for creating more accessible and inclusive digital spaces. From assistive devices to data privacy, social inclusion to sensory enhancements, this comprehensive guide provides practical solutions for individuals and organizations looking to improve the lives of vulnerable populations. With a focus on empathy, creativity, and innovation, this book is a must-read for anyone interested in harnessing the power of technology to build a more equitable and inclusive world." – Author & CEO Janet Downs

This book is dedicated to all the individuals with disabilities who have fought tirelessly for their rights, as well as those who continue to push for a more accessible and inclusive society. Your bravery, resilience, and determination inspire us every day.

Preface

The rapid development of technology has brought about unprecedented opportunities for innovation and progress, but it has also highlighted the need for greater accessibility and inclusion for people with disabilities. In this book, we aim to explore the ways in which technology can be harnessed to create more equitable and accessible digital spaces. Through discussions of assistive devices, data privacy, social inclusion, and sensory enhancements, we hope to provide insights and strategies for individuals and organizations looking to improve the lives of vulnerable populations.

Introduction

The intersection of technology and disability is a complex and multifaceted one, encompassing a wide range of challenges and opportunities. While technology has the potential to greatly improve the lives of people with disabilities, it can also perpetuate existing inequalities and exacerbate marginalization. In this book, we aim to explore the ways in which technology can be leveraged to create more inclusive and accessible digital spaces. By examining the issues and opportunities at the forefront of this intersection, we hope to offer practical solutions and insights for individuals and organizations looking to create a more equitable and just world.

CHAPTER ONE

Data as a Tool for Empowerment: Creating Inclusive Interfaces for Vulnerable Individuals

Chapter 1 of "Sensory Enhancements for a More Connected World" explores the crucial role of data in empowering vulnerable individuals through the creation of inclusive interfaces. The chapter starts with an introduction to the importance of inclusive interfaces, particularly for individuals with disabilities or limited resources.

Introduction to the importance of inclusive interfaces for vulnerable individuals

Inclusive interfaces are those that are designed with the needs of all users in mind, including those who are traditionally excluded or marginalized. This is particularly important for vulnerable individuals, who may face additional barriers to accessing and using digital technologies.

The chapter highlights how data can be used as a tool for empowerment, enabling vulnerable individuals to access information, services, and opportunities that they might otherwise be unable to access. By collecting and analyzing data on user needs, preferences, and behaviors, designers can create interfaces that are tailored to the needs of vulnerable individuals, including those with disabilities or limited resources.

Inclusive interfaces have the potential to improve the quality of life for vulnerable individuals, providing them with greater independence, autonomy, and access to resources. For example, an inclusive interface for a financial management tool could include features like voice recognition, high contrast settings, and larger fonts to accommodate users with visual impairments.

The chapter emphasizes the importance of taking a user-centered approach to designing inclusive interfaces, prioritizing the needs and preferences of vulnerable individuals. By doing so, designers can create interfaces that are not only accessible and usable but also empowering and inclusive.

Overall, Chapter 1 provides an important foundation for the rest of the book, highlighting the importance of inclusive interfaces for vulnerable individuals and the role of data in creating these interfaces. By focusing on the needs of vulnerable individuals and harnessing the power of data, designers can create interfaces that are truly empowering and inclusive, enabling all individuals to access the benefits of digital technologies.

The role of data in creating inclusive interfaces

Inclusive interfaces are designed to meet the needs of all users, including those who may be vulnerable or marginalized. Data and data response detection play a crucial role in creating these interfaces, as they enable designers to understand user needs and preferences and create interfaces that are tailored to those needs.

Data and data response detection can be collected through a

variety of means, including user feedback, surveys, and analytics. By analyzing this data, designers can gain insights into user behavior and preferences, identifying areas where interfaces can be improved to better serve vulnerable individuals.

For example, if a financial management tool is used by individuals with disabilities or limited resources, designers can collect data on the most common challenges faced by these users, such as difficulty reading small fonts or navigating complex menus. With this data and data response detection, designers can make changes to the interface that make it easier and more accessible for these users.

The chapter also highlights the importance of using data and data response detection to create interfaces that are inclusive of all users, rather than just those who are traditionally included in the design process. By collecting data from a diverse range of users, designers can create interfaces that are truly inclusive and accessible to all.

Overall, the chapter emphasizes that data and data response detection are crucial tools in creating inclusive interfaces for vulnerable individuals. By collecting and analyzing data on user needs and preferences, designers can create interfaces that are tailored to those needs, improving the quality of life for vulnerable individuals and promoting greater social justice.

LiDAR (Light Detection and Ranging) is a type of technology that uses lasers to measure distances and create detailed 3D maps of environments. LiDAR is used in a variety of applications, including autonomous vehicles, robotics, and surveying.

In terms of creating inclusive interfaces for vulnerable individuals, LiDAR can be used as a detection tool to help individuals with disabilities or limited mobility navigate their environment. For example, LiDAR can be used to detect obstacles in the path of a wheelchair user and provide real-time feedback on how to avoid them.

Similar handicapped detection gear, such as infrared sensors or ultrasonic sensors, can also be used to help individuals with disabilities navigate their environment. These sensors can detect objects in the user's path and provide feedback to the user to help them avoid obstacles.

Overall, LiDAR and other handicapped detection gear are important tools in creating inclusive interfaces for vulnerable individuals. By using these technologies to detect and navigate obstacles, individuals with disabilities or limited mobility can have greater independence and freedom of movement, promoting greater social justice and inclusion.

The potential benefits of empowering vulnerable individuals through data

Empowering vulnerable individuals through data has the potential to bring many benefits. First, it can help to promote greater independence and autonomy for individuals with disabilities or limited resources. By using data-driven interfaces, these individuals can more easily navigate their environment, access essential services, and communicate with others.

Data can also help to improve the overall quality of life for vulnerable individuals. For example, by using data to create more efficient transportation systems, individuals with disabilities or limited mobility can have greater access to work, education, and social activities. Additionally, by using data to design more accessible buildings and public spaces, individuals with disabilities can more easily access essential services and participate in their communities.

Empowering vulnerable individuals through data can also promote greater social justice and equity.

By providing access to data-driven interfaces and services, individuals who may have been excluded from traditional systems can have greater access to essential resources and opportunities. This can help to reduce inequality and promote greater diversity and inclusion in society.

Overall, there are many potential benefits to empowering vulnerable individuals through data. By leveraging the power of data to create more inclusive interfaces and services, we can promote greater independence, improve quality of life, and foster greater social justice and equity for all individuals.

Imagine a tech company that provides customer support for a popular software product. They have the ability to remotely access the user's screen to diagnose problems and provide assistance. However, in the context of Sensory Enhancements for a More Connected World, this partnership would involve a more meaningful exchange of information between the user and the Open AI system, such as ChatGPT.

For example, imagine a disabled user who relies on ChatGPT for work or entertainment. They may encounter certain barriers or difficulties in using the system that are not immediately apparent to the support team. By establishing a partnership between the user and the Open AI system, the support team could gain a more comprehensive understanding of the user's experience and the specific challenges they face.

This could involve sharing anonymized user data with the Open AI system, such as how often certain phrases or commands are used, or what types of errors or misunderstandings occur most frequently. In turn, the Open AI system could provide personalized recommendations or modifications to the interface to better meet the user's needs.

By establishing a more meaningful partnership between users and Open AI systems, we can create interfaces that are truly inclusive and responsive to the unique needs and challenges of vulnerable individuals. This can help to promote greater independence, autonomy, and quality of life for all individuals, regardless of their abilities or resources.

Examples of successful inclusive interfaces for vulnerable individuals

There are many examples of successful inclusive interfaces for vulnerable individuals. Here are a few:

Screen readers

These are software programs that read text out loud for individuals with visual impairments. They can also convert text to Braille for those who are deafblind. Screen readers have become an essential tool for many people with disabilities, allowing them to access a wide range of digital content and applications.

Janet Downs is a blind individual who relies on a screen reader to navigate her computer. She uses JAWS (Job Access With Speech), a popular screen reader software, to access email, browse the internet, and perform tasks for her job.

Speech recognition software

This technology allows individuals with mobility impairments to control their computer using their voice. By speaking commands, users can open applications, navigate the web, and even dictate text. Speech recognition software has been a game-changer for many people with disabilities, giving them greater independence and access to technology.

John Carter is a quadriplegic individual who uses speech recognition software to control his computer. He uses Dragon NaturallySpeaking, a popular speech recognition software, to compose emails, browse the internet, and perform tasks for his job.

Haptic feedback devices

These are devices that provide tactile feedback to users, such as vibrations or pressure changes. They can be used to alert individuals with hearing impairments to incoming messages or notifications, or to provide navigation assistance for individuals with visual impairments. Haptic feedback devices have also been used in gaming and virtual reality applications, providing a more immersive experience for users.

Sarah Johnson is a deafblind individual who uses a haptic feedback device to communicate with others. She uses a braille display, a device that displays text in braille and uses haptic feedback to simulate the sensation of touching physical objects, to communicate with others.

LiDAR technology

As mentioned earlier, LiDAR technology can be used to detect obstacles and provide spatial information for individuals with mobility impairments. LiDAR technology has been incorporated into a variety of devices, including canes, wheelchairs, and even drones.

Mark Thompson is a visually impaired individual who uses LiDAR technology to navigate his environment. He uses a cane with LiDAR sensors, which emits a series of laser beams that bounce off objects and provide information about his surroundings, to navigate his environment safely.

By designing interfaces with the specific needs of vulnerable individuals in mind, we can create technology that is truly inclusive and empowering. These examples demonstrate how technology can be adapted to meet the unique needs of individuals with disabilities, providing greater access, independence, and quality of life.

Chapter Two

Two-Party Interfaces: The Challenges of Creating Safe and Inclusive Digital Spaces

As digital technology becomes more ubiquitous in our daily lives, it's important to consider how it can be used to create safe and inclusive spaces for all individuals, including those who are vulnerable. In this chapter, we'll explore the concept of two-party interfaces, which refer to digital spaces that involve interactions between two parties: the user and the machine.

We'll examine the unique challenges of creating safe and inclusive two-party interfaces, particularly in the context of Open AI creations. These interfaces must be designed to provide 360-degree safety to vulnerable parties, such as individuals with disabilities or limited resources. We'll discuss the ethical implications of using data for the developmentally disabled and how this impacts individuals and society as a whole.

Furthermore, we'll delve into the concept of "battered concrete fences," which refers to individuals who are subjected to strict terms and conditions that are systemically framed in the past, without regard for their current needs or circumstances. We'll argue that people should not be subjected to these strict terms and conditions and that data should be used to improve the lives of individuals, rather than reinforcing past inequalities.

Overall, this chapter will raise important questions about the challenges of creating safe and inclusive digital spaces, as well as the need for built-in remedies and ultra-safety measures for vulnerable individuals in two-party interfaces. We'll explore how technology can be used to promote social justice and empower vulnerable individuals, rather than perpetuating past inequalities.

An exploration of the challenges that vulnerable individuals face in two-party digital interfaces

In two-party digital interfaces, vulnerable individuals often face numerous challenges that can leave them feeling marginalized and excluded. These challenges can range from accessibility issues to safety concerns and can make it difficult for them to navigate and fully participate in digital spaces.

For example, individuals with disabilities may struggle to access digital interfaces that are not designed with their specific needs in mind, such as interfaces that rely on visual cues rather than audio or haptic feedback. This can leave them feeling frustrated and excluded from the digital world, which can have a significant impact on their sense of self-worth and belonging.

Furthermore, vulnerable individuals may also face safety concerns when using two-party digital interfaces, particularly when interacting with AI-powered systems. For instance, if an individual is unable to clearly express their needs or boundaries to an AI system, they may be at risk of being exploited or manipulated. This risk can be particularly high for individuals with limited resources or those who are socially isolated.

Overall, the challenges that vulnerable individuals face in two-party digital interfaces can be complex and multifaceted. They can have a profound impact on their sense of safety, belonging, and self-worth, highlighting the urgent need for inclusive and safe digital spaces. In the following sections, we will explore these challenges in more detail and discuss potential solutions for creating safe and inclusive two-party digital interfaces.

Janet Downs is a talented software engineer who is blind but paralyzed. Despite her physical limitations, she has always been passionate about using technology to create inclusive interfaces that can benefit everyone. However, Janet often feels worried and scared when working on two-party digital interfaces with her colleagues and company customers. She fears that her work will be exploited, and her colleagues or customers might take advantage of her disability.

For example, Janet once worked on a two-party digital interface project that involved financial information exchanges. She designed the interface to have built-in remedies and ultra-safety measures to prevent vulnerable individuals from being exploited. However, during a presentation of the interface to company customers, Janet was horrified to discover that her colleagues had removed some of these safety measures, which they believed would slow down the process and make the interface less efficient. Janet felt helpless and frustrated, as her concerns were not taken seriously, and she was not given the chance to explain the importance of these safety measures.

Janet's experience highlights the challenges that vulnerable individuals face in two-party digital interfaces. It is essential to recognize and address these challenges to create safe and inclusive digital spaces that empower everyone, regardless of their abilities or limitations.

The ethical considerations of creating safe and inclusive digital spaces

Creating safe and inclusive digital spaces is not just a technical challenge, but also an ethical one. It is important to consider the rights of vulnerable individuals and ensure that their privacy and safety are not compromised in any way.

One ethical consideration is the issue of consent. Vulnerable individuals should be able to give informed consent before sharing any personal information or participating in any online activity. It is important to explain the purpose of the digital space and how their personal information will be used. Consent should be obtained in a way that is accessible and easy to understand.

Another ethical consideration is the issue of bias. Technology is not neutral, and biases can be embedded in algorithms, data sets, and interfaces. It is important to be aware of these biases and work to eliminate them, so that vulnerable individuals are not unfairly discriminated against.

Furthermore, transparency and accountability are essential in creating safe and inclusive digital spaces. Companies and organizations should be transparent about their data collection practices and how they use the data. They should also be accountable for any harms caused to vulnerable individuals as a result of their digital interfaces.

Lastly, there is the issue of power dynamics. Vulnerable individuals may be at a disadvantage in two-party digital interfaces, as they may have less power to control the interactions. It is important to consider how power dynamics may affect the safety and inclusivity of the digital space, and work to empower vulnerable individuals.

One example of when ethical considerations were neglected is the case of Cambridge Analytica, a political consulting firm that harvested the personal data of millions of Facebook users without their informed consent. This data was then used to influence the 2016 US Presidential election and the Brexit vote in the UK.

In this case, individuals were not made aware that their data was being collected and used for political purposes, violating the principle of informed consent. Additionally, the algorithms used by Cambridge Analytica were found to be biased, with the company targeting specific groups with personalized messages designed to sway their opinions. The lack of transparency in how this data was collected and used also contributed to the ethical concerns, as individuals were not aware of the extent to which their personal information was being exploited.

Finally, when the scandal broke, there was a lack of accountability for those responsible for the breach, with the company shutting down and its executives avoiding prosecution. This lack of consequences for unethical behavior sends a message that such actions are acceptable, which further erodes trust in the digital world and hinders efforts to create safe and inclusive digital spaces.

Overall, creating safe and inclusive digital spaces requires a deep understanding of the ethical considerations at play. It is not just about creating accessible interfaces, but also about respecting the rights and dignity of vulnerable individuals.

The importance of designing interfaces with the user's needs in mind

When designing interfaces for vulnerable individuals, it is crucial to keep their needs and abilities in mind. One-size-fits-all solutions often fall short of meeting the needs of diverse user groups, including those with disabilities or limited digital literacy.

Designers must strive to create interfaces that are intuitive, easy to navigate, and tailored to the user's unique needs. This requires a deep understanding of the user's perspective and the challenges they face.

For example, a person with limited mobility may require a simplified interface with larger buttons and fewer options. Similarly, someone with hearing impairment may need captions or audio descriptions to fully access the content.

One recent example of an ethical consideration being considered and benefiting everyone is the implementation of two-factor authentication for online accounts. This added layer of security helps protect users' personal information and prevent unauthorized access to their accounts. While it may be seen as an inconvenience for some users, the overall benefit of increased security far outweighs any temporary inconvenience. By prioritizing the users' security needs and implementing this measure, both the users and the companies providing the online services can benefit from increased trust and protection.

Jane is a visually impaired individual who uses a screen reader to navigate the web. She frequently orders groceries online, but often struggles to find the nutritional information for each product. Jane reached out to the grocery store's customer support team, requesting that they make the nutritional information more accessible for individuals with visual impairments. The grocery store took Jane's feedback into consideration and worked with their web development team to implement an audio feature that reads out the nutritional information when Jane hovers over each product. This not only benefited Jane, but also other visually impaired individuals who shop at the grocery store, making their shopping experience more inclusive and empowering.

Designers must also consider the potential barriers that certain design choices may create for some users. For instance, the use of color-coding to convey information may exclude individuals with color blindness.

By putting the user's needs at the forefront of design decisions, designers can create interfaces that are more inclusive, accessible, and effective. Such interfaces can not only empower vulnerable individuals but also enhance the user experience for all users, regardless of their abilities.

Examples of inclusive interfaces that prioritize user ultra-safety

In creating inclusive interfaces, it is important to prioritize user safety, especially for vulnerable individuals. One example of an inclusive interface that prioritizes user safety is the app "Be My Eyes," which was developed to assist people who are visually impaired. The app connects users to a network of volunteers who can provide visual assistance through live video calls. The app's interface is designed to ensure that the volunteer cannot see or access any of the user's personal information, including their name, location, or contact information. This ensures that the user's safety and privacy are protected.

Another example is the "Emergency Chat" feature on the Uber app, which allows riders to communicate with their driver without sharing their phone number or other personal information. This is particularly important for individuals who may be in vulnerable situations, such as women traveling alone at night. The feature ensures that riders can communicate their needs and concerns to their driver without compromising their safety or privacy.

In the healthcare industry, there are also examples of inclusive interfaces that prioritize user safety. For instance, the "MedSnap ID" app uses image recognition technology to identify prescription drugs and ensure that patients receive the correct medication. This helps to prevent medication errors that can have serious consequences for patient safety.

The "Pain Squad" app is an excellent example of an inclusive interface that prioritizes user ultra-safety. This app was designed to help children with cancer and their doctors detect pain in real-time, so that they can receive immediate treatment. The app is designed like a game and has different levels, where children become part of a team of detectives investigating crimes. As they play, they are asked about their pain levels, and the data is sent to the doctor, who can then provide the necessary treatment. This app helps to empower children with cancer to be more involved in their treatment process and provides doctors with vital information to make informed decisions. The app is an excellent example of an inclusive interface that considers the needs of vulnerable individuals, and prioritizes user safety.

Suppose there is a person named Adam who suffers from chronic pain due to a medical condition. In the past, Adam had to endure the lag time between when he feels the pain and when he receives effective treatment. However, with the advent of wearable devices that can monitor vital signs and send real-time alerts to medical professionals, Adam no longer has to worry about the lag time.
The device can immediately detect when Adam is experiencing severe pain and alert his doctor, who can then take appropriate measures to alleviate his pain. This ensures

that Adam receives prompt and effective treatment, making his life much easier and less painful.

Overall, these examples demonstrate the importance of designing interfaces that prioritize user safety and privacy. By doing so, we can create inclusive digital spaces that empower vulnerable individuals and promote their well-being.

Chapter Three

Battered Concrete Fences: Overcoming Systemic Barriers to Inclusive Interfaces

As we delve into Chapter 3, we begin to address some of the more daunting challenges facing the creation of inclusive interfaces.

While we have explored the importance of designing with the user's needs in mind and the ethical considerations that must be taken into account, there are still systemic barriers that prevent vulnerable individuals from accessing the digital world on an equal footing with the rest of society. These barriers are like battered concrete fences that have been put up over time and seem insurmountable. However, by taking a closer look and breaking down these barriers, we can identify strategies for overcoming them and creating a more inclusive digital world. Let's explore the systemic barriers and how we can work together to overcome them.

A critical examination of the systemic barriers that prevent vulnerable individuals from accessing safe and inclusive interfaces

When it comes to creating safe and inclusive interfaces for vulnerable individuals, it's important to acknowledge that systemic barriers can often get in the way. These barriers can take many forms, from inadequate funding for assistive technology to discriminatory practices in the workplace. For example, a person with a disability may struggle to access digital interfaces because of a lack of captioning or audio descriptions. Or they may encounter physical barriers to access, such as a lack of ramps or elevators in a building. Additionally, there may be social and cultural barriers that prevent certain groups of people from feeling comfortable using technology, such as language barriers or fears of discrimination. It's crucial that we identify and address these systemic barriers in order to create truly inclusive digital spaces.

Imagine a person named Sarah, who uses a wheelchair due to a spinal cord injury. Sarah works as a freelance writer and often has to attend online meetings with clients and colleagues. However, she struggles to access some of the virtual meeting platforms because they are not designed to be accessible for people with disabilities. On one occasion, Sarah was unable to attend an important meeting with a potential client because the platform they were using did not provide closed captioning or any other accessibility features. This resulted in a loss of business and a feeling of frustration and disappointment. Sarah's story is just one example of how systemic barriers can prevent vulnerable individuals from accessing safe and inclusive interfaces.

Emily is a 9-year-old girl with Down syndrome who attends a public school. Her mom, Lisa, noticed that Emily was becoming increasingly withdrawn and disinterested in school. Lisa tried to talk to Emily's teacher, but the teacher didn't seem to have any solutions or strategies to help. It wasn't until

Lisa had a meeting with the school's special education coordinator that she realized that the real problem was with the school's physical environment.

Emily's classroom was located on the second floor of the building, and there was no elevator or ramp to access it. Emily had to climb a long set of stairs every day, which was exhausting and intimidating for her. Lisa realized that this was likely the reason why Emily was so unhappy at school - she was too overwhelmed by the physical barriers to even focus on her learning.

Lisa and the special education coordinator worked together to find a solution. They advocated for the installation of a ramp or elevator in the school building, which would make it much easier for Emily to access her classroom. It took some time, but eventually, the school administration agreed to make the necessary changes. With the new accessibility measures in place, Emily was able to access her classroom with ease, and Lisa noticed a significant improvement in her daughter's demeanor and engagement with school.

In the past, there were instances where those in power, who had access to vast amounts of data, would treat those who were at the end of the data chain with disregard and even exploitation. This often happened because the individuals who were most impacted by the data had the least power to influence how it was collected, analyzed, and used. They were often seen as mere data points, and not as real people with unique experiences and needs. As a result, their voices were ignored, and their concerns were not taken into account when making decisions that impacted their lives. This led to a situation where vulnerable individuals were left behind, and their needs were not adequately addressed by the design of digital interfaces.

Just like how certain resources are necessary to sustain oneself, some individuals in positions of authority today have access to vast amounts of information. They use this information to make decisions that affect others, but often fail to take into account the impact of their choices on those at the bottom of the totem pole. These individuals are frequently viewed merely as numbers or statistics, rather than unique individuals with their own emotions and thoughts. Consequently, their concerns and input are typically disregarded, and they have no say in how their data is collected or used. This creates a situation where vulnerable individuals are excluded, and their requirements are not considered in the development of digital tools. A common example of this is the usage of personal data by large tech companies without adequate consent or transparency, leading to privacy violations and other adverse effects on individuals.

In June 2020, a 20-year-old Robinhood trader named Alexander Kearns started trading options on the Robinhood app. Kearns had only been trading for a few months and had started with a small investment of $5,000, which he had grown to around $16,000. However, when he checked his Robinhood account one day, he saw a negative $730,000 balance. Kearns panicked and sent a message to Robinhood's customer support, but he received an automated response telling him that it might take a few days to get back to him. Later that day, Kearns took his own life. It was later revealed that the negative balance was caused by a glitch in the app's interface that incorrectly displayed the value of Kearns' trades. The incident raised questions about the design of the app and the responsibility of financial technology companies to ensure the accuracy and clarity of their interfaces.

Everyday glitches like the Robinhood app user's negative balance can have serious implications for people with visual, PTSD, or developmental disorders. For example, individuals with visual impairments may rely on digital tools to navigate their daily lives, and glitches or errors in these tools can lead to confusion, frustration, or even danger. Those with PTSD may be triggered by unexpected or erroneous information, leading to distress and further exacerbation of their symptoms. Similarly, individuals with developmental disorders may struggle to process and understand complex or ambiguous information, which can cause anxiety and distress. In general, these glitches can create additional barriers for vulnerable individuals and make it even more difficult for them to access and benefit from digital tools and interfaces.

The importance of addressing past inequalities in data collection and analysis - rescue operations to save lives

The consequences of past inequalities in Open AI data reflections can be a matter of life or death. If vulnerable communities are excluded from data collection and analysis, they may not receive adequate rescue operations in emergency situations. For example, imagine a flood hitting a neighborhood where many residents speak a language that is not commonly spoken in the area. If this community's language is not included in Open AI data reflections, emergency responders may not be able to communicate with them effectively and provide the necessary assistance. This is just one example of how past inequalities in Open AI data reflections can have dire consequences in emergency situations.

Open AI data reflections have been a source of past inequalities that continue to impact people's lives today. Imagine a situation where a medical diagnosis is made based on incomplete or biased data. For example, if a study on heart disease only includes data from white men, it may miss important factors that affect women or people of color. This could result in misdiagnosis or delayed treatment, which could have life-threatening consequences. In a more elementary example, imagine a teacher only evaluating their students based on one test or assignment without taking into account individual learning styles or circumstances. This could lead to unfair grades and hinder a student's academic progress. By addressing past inequalities in Open AI data reflections, we can ensure that everyone's unique experiences and needs are taken into account and ultimately save lives in critical situations.

Past inequalities in data collection and analysis have led to exclusion and marginalization of vulnerable groups, which can have dire consequences when it comes to saving lives. For example, during natural disasters, emergency responders may rely on Open AI data reflections to locate survivors and coordinate rescue efforts. But if the data is incomplete or biased towards certain groups, some individuals may be left behind, putting their lives at risk. A more elementary example could be a medical emergency where a patient's medical history is incomplete or inaccurate due to past inequalities in data collection, leading to incorrect treatment and potentially fatal consequences. It is therefore essential to address these past inequalities in Open AI data reflections to ensure that everyone, especially vulnerable groups, are included and considered in rescue operations.

Safer private networks for surviving systemic barriers and exclusive interfaces

Creating a safe and private network is crucial in overcoming systemic barriers and exclusive interfaces. When vulnerable individuals and marginalized communities are forced to rely on public networks that do not prioritize their safety, they risk being exposed to harmful content or attacks. As such, it is essential to design private networks that can protect users from these dangers.

A safe private network ensures that users have control over their data reflections, pre-future uses and are free from any forms of intrusion or exploitation. By building secure channels of communication and data transfer, individuals can connect and interact with others without the fear of being exposed to potential harm. Such networks can also provide greater anonymity and privacy, allowing individuals to share information and ideas freely without fear of retribution.

An excellent example of a safer private network is the Tor browser, which enables users to access the internet while maintaining their anonymity. This browser helps individuals evade surveillance and censorship, especially in countries where freedom of expression is not guaranteed. Another example is Signal, a private messaging app that uses end-to-end encryption to protect users' messages from being intercepted by unauthorized third parties.

Imagine a parent who sets up a safe and secure private network for their child to use while accessing online resources. By doing so, the parent ensures that their child's personal information and online activities are protected from malicious individuals who may seek to exploit or harm the child. Additionally, the child is able to learn and explore the online world in a safe and controlled environment, free from inappropriate content or harmful interactions. This not only benefits the child and the parent, but also benefits the larger community by promoting safe and responsible online behavior for all users.

"Lock-out, tag-out" is a common practice in industries where workers come into contact with dangerous machinery or equipment. This practice involves physically locking the machinery or equipment in question to prevent accidental or unauthorized use, with a tag clearly indicating the reason for the lock-out. Similarly, with Open AI data, certain sensitive data or algorithms may require special precautions to prevent unauthorized access or unintended use. By using secure networks and implementing strict access control measures, companies can ensure that their data and algorithms are protected from misuse or unauthorized access, much like how "lock-out, tag-out" helps prevent accidents in the workplace. Just like the importance of following proper procedures in the workplace, it's important for organizations to prioritize the security of their Open AI data to prevent potential harm to users or the public.

Parents or caretakers of vulnerable and handicapped individuals often have an intimate understanding of their specific needs, including what data needs to be protected or locked out. This is because each individual has unique vulnerabilities and sensitivities that require tailored support and protection. For example, a child with a visual impairment may have a heightened sensitivity to flashing lights or certain colors, which could trigger seizures. A parent or caretaker who is intimately familiar with the child's needs and sensitivities would be in the best position to identify potential risks and ensure that the child's access to data is restricted or tailored appropriately. Similarly, for individuals with PTSD or developmental disorders, caretakers can help identify triggers and ensure that sensitive information or experiences are not inadvertently exposed. In this way, parents and caretakers play a critical role in protecting vulnerable individuals from potential harm and ensuring that their digital experiences are safe and inclusive.

On the flip side, consider a scenario where a pharmaceutical company releases a new drug to the market without fully testing it on vulnerable populations such as pregnant women or elderly individuals. The drug is later found to have adverse effects on these populations, causing harm and even death.

This failure to ethically consider the most vulnerable populations in the drug's development and testing process can have devastating consequences and violates the principle of non-maleficence, which is an essential ethical principle in healthcare. Such actions can also erode trust in the healthcare system and harm future research efforts. It is critical that vulnerable populations are adequately represented and protected in all stages of research and development to ensure their safety and well-being.

In conclusion, creating safe private networks can provide a lifeline for individuals who have been chronically battered by systemic barriers and exclusive interfaces. By ensuring the safety and privacy of vulnerable individuals, we can build a more inclusive and equitable digital world that prioritizes the well-being and agency of all its users.

CHAPTER FOUR

Self-Reflective Detection Data for Social Justice: The Ethics of Creating Two-Party Interfaces

Chapter 4 delves into the ethical considerations of creating two-party interfaces, and how self-reflective detection data can be used to promote social justice. As technology advances and data collection becomes more pervasive, it is important to consider the impact that these interfaces can have on marginalized populations. This chapter will explore the ways in which these interfaces can perpetuate systemic biases and inequalities, as well as how they can be designed to promote inclusivity and equity. The creation of two-party interfaces must be done with an awareness of power dynamics and an understanding of the complex ways in which data can be used to reinforce or challenge existing systems of oppression. By centering the needs and experiences of those who are most vulnerable, self-reflective detection data can be used to create more just and equitable interfaces.

A deep dive into the ethical considerations of using reflective data to create safe and inclusive digital spaces

Ah, it's time to get all dressed up and dive into the exciting world of reflective data ethics! Just like putting on a great outfit for a big event, creating safe and inclusive digital spaces requires careful consideration and attention to detail.

Reflective data is a powerful tool for understanding how people interact with digital interfaces, but it also comes with important ethical considerations. As we design custom interfaces that use this data to create safer and more inclusive spaces, we must be mindful of how it might be used to exploit vulnerable individuals.

For example, let's say we are accessing a dating app that uses reflective data to improve matches and create a more positive user experience. While this data can be helpful in understanding what users are looking for and what they enjoy, it can also be used to manipulate individuals into staying on the app for longer periods of time or making purchases they wouldn't otherwise make.

To avoid these ethical pitfalls, we must design interfaces that prioritize user privacy and informed consent in real-time. This means being co-transparent and ongoing about what data is being co-collected, how it is being used, and giving producers and users the ability to opt-out or delete their data at any time.

Imagine a person with a disability who uses a wheelchair to move around. They visit a website that offers an accessible interface for individuals with disabilities. The website informs them about what data it collects, why it collects it, and how it will be used. The user is given the option to consent to the collection and use of their data or decline it. The website also provides clear instructions on how the user can delete their data whenever they want. The interface is designed in such a way that the user feels comfortable and safe to use it, knowing that their privacy is respected and their data is being used in a transparent and ethical manner.

Imagine if ChatGPT and its Open AI affiliates allowed users to provide feedback in real-time and actively incorporated that feedback into their data collection and analysis. This would create a more two-way conversation, where users could express their needs and concerns and producers could respond accordingly. For example, if a user with dyslexia found it difficult to read certain fonts, they could provide feedback and suggest alternative fonts that are more accessible for them. Producers could then incorporate this feedback into their design choices, making the interface more inclusive for all users with similar needs. This would ensure that the user experience is constantly improving and evolving to meet the needs of all users, including those with custom vulnerabilities.

Top-down data collection can often feel invasive and disempowering to vulnerable people, as they may not fully understand how their data is being collected and used. It may also not take into account their unique needs and experiences, leading to exclusion and marginalization.

On the other hand, a full-circle, privately inclusive approach to data collection and analysis can make vulnerable people feel more empowered and included. For example, imagine a healthcare provider using a comprehensive, inclusive approach to data collection and analysis for their patients with disabilities. The provider engages with patients and their families to understand their unique needs and experiences, and takes steps to ensure that data is collected in a way that is respectful and inclusive of their privacy and autonomy. This approach can lead to more accurate and personalized care, as well as a sense of trust and partnership between the provider and the patient.

By prioritizing these ethical considerations, we can create digital spaces that are both safe and inclusive for all users, regardless of their backgrounds or vulnerabilities. So let's get dressed up and dive into the exciting world of reflective data ethics!

The role of co-technology in promoting social justice

Co-technology, also known as collaborative technology, refers to the use of technology to promote collaboration and cooperation among individuals and groups. In the context of social justice, co-technology plays a crucial role in promoting equality, inclusivity, and fairness.

Think of co-technology like a superhero team-up - just like how Batman and Robin or the Avengers work together to fight evil and save the world, co-technology brings people together to address social injustices and make the world a better place.

One example of co-technology in action is the use of online platforms to facilitate communication and collaboration between individuals and organizations working towards social justice. For instance, a platform like Change.org enables people to start petitions and mobilize others to support their cause. Another example is the use of social media to raise awareness about issues and organize protests and rallies.

Another example of co-technology is when a group of people collaborate to create a project using a shared document or online platform, such as Google Docs or Slack. Each member has equal access to the document and can contribute to it in real-time, allowing for a collaborative effort that values and includes everyone's input. In this way, co-technology promotes social justice by providing an inclusive space where everyone's ideas are heard and valued, regardless of their background or position.

Co-technology also involves the use of innovative solutions to address social problems. For example, a group of engineers and designers might work together to create affordable and accessible assistive technology for people with disabilities. This collaboration not only results in a tangible product that can improve the lives of those with disabilities but also helps to break down the barriers and stigma that these individuals often face.

The use of co-technology can also create new opportunities for underrepresented communities to have their voices heard. For instance, an online forum or community platform can provide a safe and inclusive space for marginalized groups to connect with one another and share their experiences. This can lead to greater understanding and empathy, and ultimately, drive positive change.

In short, co-technology plays a crucial role in promoting social justice by bringing people together, providing innovative solutions to problems, and creating new opportunities for marginalized communities to have their voices heard.

The potential consequences of failing to prioritize Sensory Enhancements for a More Connected World in two-party interfaces

As we strive towards creating more inclusive and accessible technology, it is important to consider the potential consequences of failing to prioritize Sensory Enhancements for a More Connected World in two-party interfaces. When data is collected without taking into account the needs and perspectives of disabled individuals, it can lead to further marginalization and exclusion.

For example, let's say a company creates a new fitness tracking app that relies heavily on data collected from wearable devices. If the app is not designed with accessibility in mind, it may exclude individuals with disabilities who are unable to use the wearable device or access the data in a meaningful way. This exclusion can lead to feelings of frustration, isolation, and a sense of being left out of important health and wellness conversations.

Moreover, failure to prioritize Sensory Enhancements for a More Connected World in two-party interfaces can result in more than just exclusion from technology. It can also have severe consequences on disabled individuals' ability to navigate and participate in society. For instance, if a transportation app fails to consider the needs of wheelchair users, they may be unable to access critical information about accessible transportation options, limiting their ability to move around and participate in everyday activities.

If the Robinhood user or any other invisible disabled community users had access to Sensory Enhancements for a More

Connected World in two-party interfaces, it could have made a significant difference in their ability to safely and effectively use these platforms. For example, if Robinhood had taken into account the needs of users with cognitive or visual disabilities, they may have designed their interface in a way that made it easier for those users to navigate and understand. This could have prevented the tragic loss of life that occurred.

In general, prioritizing Sensory Enhancements for a More Connected World in two-party interfaces has the potential to create a more equitable and accessible online world for everyone, including those with disabilities. It could lead to the development of more inclusive platforms and technologies that empower all users to participate in the digital economy and society. By failing to prioritize Sensory Enhancements for a More Connected World, we risk excluding an entire segment of the population from the benefits of the digital age, perpetuating inequalities, and denying them access to essential resources and opportunities.

By prioritizing Sensory Enhancements for a More Connected World in two-party interfaces, we can create a more equitable and accessible future for all individuals, regardless of their abilities. This means considering the unique needs and perspectives of disabled individuals throughout the entire design and development process, and making necessary adjustments to ensure that everyone can participate in and benefit from technology advancements.

Best practices for designing proven interfaces with social justice in mind

When designing interfaces that prioritize social justice and inclusivity, it's important to consider a few key best practices. Here are some suggestions:

1. Prioritize co-design and co-development: Including diverse voices in the design and development process can help ensure that the interface is accessible and inclusive for all users. This can include

consulting with community members, advocates, and people with disabilities to gather feedback on design decisions.

2. Ensure transparency and informed consent: Users should always be aware of what data is being collected and how it is being used. It's important to clearly explain the purpose of the interface and the data that is being collected. Additionally, users should have the ability to opt-out or delete their data at any time.

3. Test for accessibility and inclusivity: Interfaces should be tested for accessibility and inclusivity throughout the development process. This includes testing for screen readers, keyboard-only navigation, and color contrast, as well as ensuring that the interface is easy to use for people with a range of abilities and disabilities.

4. Consider intersectionality: It's important to recognize that different communities may face multiple

barriers to accessing technology. Interfaces should be designed with intersectionality in mind, taking into account the ways that different forms of oppression and marginalization intersect and compound.

5. Engage in ongoing evaluation and improvement: Technology is constantly evolving, and interfaces should be regularly evaluated and improved upon. This includes gathering feedback from users and engaging in

ongoing accessibility testing and improvements.

During the pre-design of the Pain Squad app, great attention was given to the most vulnerable first, including children and adolescents with cancer who experience pain regularly. The developers of the app worked closely with patients, families, and healthcare professionals to ensure that the app was designed with their needs and preferences in mind. This involved conducting user testing and obtaining feedback at each stage of the development process to ensure that the app was user-friendly and met the unique needs of this vulnerable population.

The app was designed to be engaging and interactive, using game-like elements to encourage children and adolescents to regularly report their pain levels and symptoms. This allowed healthcare providers to monitor pain and symptom management over time and make necessary adjustments to treatment plans.

By prioritizing the needs of the most vulnerable population in the pre-design phase, the Pain

Squad app was able to effectively address a gap in pain management for children and adolescents with cancer. This approach to design can be applied to other digital products and services, including those intended to promote social justice, to ensure that the needs of the most vulnerable are addressed in a proactive and effective manner.

As mentioned before, lock-out, tag-out procedures are designed to ensure that People are safe when working around machinery and equipment. The purpose of these procedures is to prevent accidental injury or death by ensuring that the gear is shut down and de-energized before any work is done. Similarly, before data is turned into software or digital interfaces, it is important to consider how it could potentially harm or exclude the most vulnerable members of society.

If this consideration is not given, it could be likened to not following lock-out, tag-out procedures before working on machinery. Just as failing to follow lock-out, tag-out procedures could lead to injury or death, failing to consider the potential consequences of digitizing data without ensuring that it is handicapped proof could result in harmful and exclusionary digital interfaces. Therefore, it is important to approach the design of digital interfaces with the same level of care and attention to detail as one would approach the safety procedures for working with machinery.

By following these best practices, designers and developers can help save lives and ensure that their interfaces promote social justice and inclusivity for all users.

In conclusion, the development of two-party interfaces that prioritize social justice and inclusivity requires intentional and ongoing consideration of the needs and vulnerabilities of the most marginalized populations. The potential consequences of failing to prioritize Sensory Enhancements for a More Connected World can be dire, as seen in the examples of the Robinhood user and the invisible disabled community. Therefore, it is essential to design proven interfaces with social justice in mind, incorporating best practices such as involving the most vulnerable in pre-design, utilizing co-technology, and implementing built-in remedies. Chapter 5 will explore the importance of designing for safety proofing two-party Open AI creations and discuss how built-in remedies can contribute to the creation of more accessible and inclusive digital spaces.

Chapter Five

Preventative Protocols: Designing for Safety Proofing Two-Party Open AI Co-Creations

Chapter 5's focus on designing for safety proofing two-party Open AI co-creations is essential for anyone looking to develop and deploy digital systems that prioritize user safety and privacy. This is especially important in fields such as healthcare, finance, and education, where sensitive information and personal data are involved. It is important to prioritize safety proofing from the earliest stages of development, so that potential issues can be addressed and mitigated before the system goes live. Additionally, this approach can help build trust with users, who may be more willing to engage with a system that they feel is secure and transparent. Ultimately, by implementing preventative protocols and designing for safety, we can create digital systems that are not only effective, but also ethical and socially responsible.

As we continue to explore the world of two-party open AI co-creations, it is important to ensure that we are doing so in a safe and responsible manner. Building in preventative protocols to our design process can help to safeguard against potential risks and ensure the safety and well-being of all users. In this chapter, we will take a deep dive into the world of safety proofing and explore the various measures that can be taken to create secure and protected co-creations. From designing with privacy in mind, to implementing built-in remedies for potential hazards, we will examine the best practices for creating safe and responsible two-party open AI creations. Join us as we embark on this adventure into the world of safety proofing, and discover the exciting potential of preventative protocols in open AI co-creations.

An exploration of the need for Preventative Protocols and ultra-safety measures in two-party interfaces

The world of technology is rapidly advancing, and with it comes a plethora of new opportunities and challenges. While it is certainly exciting to see the advancements in AI and co-technology, it is also important to acknowledge the potential risks and dangers that come with it. As such, it is critical to explore the need for preventative protocols and ultra-safety measures in two-party interfaces.

The traditional model of software development has often focused on reactive measures, meaning that security and safety features are added only after a breach or attack has occurred. This approach, however, has proven to be insufficient as it leaves the system vulnerable to attacks and potentially catastrophic consequences. In order to effectively ensure the safety and security of two-party interfaces, it is necessary to shift towards a more proactive approach that includes preventative protocols and ultra-safety measures.

Preventative protocols refer to the various measures taken during the design and development stages to identify and mitigate potential risks and vulnerabilities. This includes conducting thorough risk assessments, implementing secure coding practices, and conducting regular security audits. Ultra-safety measures, on the other hand, refer to the various mechanisms put in place to ensure the continued safety and security of the system. This includes features such as data encryption, multi-factor authentication, and intrusion detection systems.

The need for reflective data measures in two-party interfaces is also critical because of the potential unfair risks that come with co-creation between humans and machines. The emergence of Open AI and other AI-related technologies has presented new challenges and opportunities for innovation. However, as Sensory Enhancements for a More Connected World moves towards co-creating with AI systems, we also run the risk of unintended consequences that may harm individuals or entire communities. For instance, imagine a situation where an AI system is designed to make healthcare recommendations, but the recommendations end up being biased against certain groups of people, resulting in inadequate or harmful treatment. To avoid such scenarios, we need to explore and develop preventative protocols and ultra-safety reflective measures that can help to minimize risks, protect vulnerable populations, and ensure that co-creation with AI is safe and beneficial for all.

The need for preventative protocols and ultra-safety measures is especially critical in ulterior two-party interfaces as they involve the interaction of two distinct parties with varying levels of access and privilege. This means that any vulnerabilities in the system can be exploited by either party, potentially leading to significant harm or damage.

Let's take the example of a bank's online system, where customers can access their accounts and perform transactions. The bank has different levels of access and privilege for its employees and customers. If the system is not designed with preventative protocols and ultra-safety measures, a customer could potentially exploit a vulnerability to gain unauthorized access to an employee's account or sensitive information, such as login credentials or financial data. Similarly, an employee could exploit a vulnerability to gain unauthorized access to a customer's account or personal information. The consequences could be significant, ranging from financial loss to identity theft and reputational damage for the bank. Therefore, preventative protocols and ultra-safety measures are crucial in two-party interfaces like this to prevent such incidents from occurring.

Overall, the need for preventative protocols in two-party interfaces, internally and out, cannot be overstated. As technology continues to advance, it is critical that we prioritize ultra-safety and handicapped security in our design and development processes in order to prevent potential disasters and ensure that our technology serves us in a positive and productive manner.

Examples of Preventative Protocols that prioritize user safety

Preventative protocols are essential for ensuring the safety and security of external two-party interfaces. In this section, we will explore some examples of preventative protocols that prioritize user safety:

1. Multi-Factor Authentication: Multi-factor authentication is a security measure that requires users to provide two or more forms of identification before gaining access to a system. This could include something like a password and a fingerprint scan or a password and a security token. By requiring multiple forms of identification, multi-factor authentication adds an extra

layer of security, making it more difficult for attackers to gain unauthorized access to the system.

2. User Permissions: User permissions are a way of controlling what actions different users can perform within a system. By assigning different levels of permission to different users, the system can ensure that only authorized users are able to perform certain actions. For example, a user with limited permission may be able to view data but not

edit it, while an administrator may be able to both view and edit data. This can help prevent accidental or intentional harm to the system or its users.

3. Regular System Updates: Regular system updates are essential for maintaining the security of a system. Updates often contain security patches that address known vulnerabilities, making it more difficult for attackers to exploit them. By regularly updating the system, developers can stay

ahead of potential threats, making the system more secure and less vulnerable to attack.

4. Data Encryption: Data encryption is a process of converting sensitive information into an unreadable format, known as ciphertext. This makes the data useless to anyone who doesn't have the key to decrypt it. By encrypting sensitive data, developers can protect it from unauthorized access, making it more difficult

for attackers to steal or manipulate the data.

5. Regular Security Audits: Regular security audits are an essential part of maintaining the security of a system. Security audits involve testing the system for vulnerabilities and identifying potential threats. By regularly performing security audits, developers can identify and address vulnerabilities before they can be exploited by attackers. This can help prevent data breaches, system downtime,

and other potential threats to user safety.

Internal two-party interfaces: safety measures for everyday life

When we think of two-party interfaces, we often think of external systems like online platforms or digital communication tools. However, there are also internal two-party interfaces that we encounter in our everyday lives. These are systems where our own physical or mental abilities interact with the environment around us, whether it's getting dressed, driving a car, or handling household appliances.

For people with disabilities, these internal two-party interfaces can present unique safety challenges. For example, someone with limited mobility may struggle to safely transfer themselves in and out of a wheelchair, while someone with a visual impairment may need assistance in navigating unfamiliar environments. In these situations, it is important to consider safety measures that can help prevent accidents or injuries.

One example of a safety measure for internal two-party interfaces is the use of adaptive equipment. This can include wheelchairs with brakes, hearing aids, and other assistive devices that make it easier for people with disabilities to navigate their environments safely. Another example is the use of braille on everyday objects like elevator buttons or medication bottles, which can help people with visual impairments navigate their surroundings with greater confidence.

Childproof caps on medication bottles are another example

of safety measures for internal two-party interfaces. These caps are designed to be difficult for children to open, reducing the risk of accidental ingestion of medication. Similarly, gloves are often used as a safety measure in situations where hands need to be protected from sharp objects or hazardous materials, such as in construction work or laboratory settings.

In conclusion, safety measures for internal two-party interfaces are just as important as those for external two-party interfaces. By considering the unique safety challenges presented by internal two-party interfaces and implementing appropriate safety measures, we can help prevent accidents and injuries and ensure that everyone can navigate their environment safely and with greater confidence.

Strategies for dual-designed interfaces that are responsive to disabled users needs and prevent unnecessary harm

When designing quad-party interfaces, it is essential to consider the needs of disabled users and prioritize their safety. Here are some strategies for designing dual-designed interfaces that are responsive to disabled users' needs and prevent unnecessary harm:

-Prioritize accessibility

Design interfaces with accessibility in mind from the very beginning. For example, a website should be designed to be accessible to users with visual impairments by using proper HTML tags and providing alt text for images.

Personal example: When designing a website for a quad-party audience, consider using large, easy-to-read fonts and high-contrast color schemes to improve accessibility for users with visual impairments.

-Incorporate user feedback

Incorporate user feedback into the design process to ensure that the interface is meeting the needs of all users. This feedback can come from disabled users themselves or from disability organizations that work with quad-party communities.

Personal example: When designing a quad-party interface, conduct user testing and solicit feedback from disabled users to identify any accessibility barriers or areas for improvement.

-Conduct accessibility audits

Regularly conduct accessibility audits to identify and address any barriers to accessibility. These audits should be conducted by qualified accessibility experts who can identify issues that may not be immediately obvious to designers or developers.

Personal example: Conduct regular accessibility audits of a quad-party interface to identify any accessibility issues and implement necessary changes to improve accessibility for all users.

-Invest in training

Provide training for designers and developers to ensure that they are familiar with accessibility best practices and can incorporate them into the design process.

Personal example: Provide training for designers and developers working on a quad-party interface to ensure they are aware of accessibility best practices and can incorporate them into the design process.

-Collaborate with disabled people

Collaborate with disabled people and quad-party disability organizations to ensure that their needs and perspectives are represented in the design process.

Personal example: Partner with disability organizations and quad-party communities when designing a quad-party interface to ensure that their needs and perspectives are considered and incorporated into the final product.

-Emphasize universal design

Design interfaces with universal design principles in mind to ensure that they are accessible to all users, regardless of ability.

Personal example: Use universal design principles when designing a quad-party interface to ensure that it is accessible to all users, regardless of ability.

What to expect

As a summary, the implementation of adaptive preventative protocols can be a lifesaver for disabled people who feel out of options. By taking a quad-party approach, designing interfaces that prioritize disabled users' needs, utilizing technology and partnering with advocacy organizations, the outcomes can be transformational. For example, an individual who previously felt excluded from participating in a quad-party environment due to inaccessible technology may now be able to participate fully, thus creating more inclusive and equitable interactions. The use of adaptive technologies, such as voice recognition software or screen readers, can further support this process. Additionally, working with advocacy organizations can provide valuable insights and feedback from the disabled community, ensuring that the designs are not only accessible but also responsive to the needs of disabled users. Ultimately, prioritizing Sensory Enhancements for a More Connected World in the design and development of quad-party interfaces can have far-reaching positive effects, enabling disabled people to fully participate in society and minimizing the risk of harm or exclusion.

The story above highlights the importance of designing inclusive interfaces with adaptive preventative protocols that can respond to the needs of all users, including those with disabilities. As technology continues to advance and the world becomes increasingly interconnected, it is critical that we prioritize the safety and well-being of all individuals. This is especially true for vulnerable populations who may have limited options and face significant challenges in accessing the tools and resources they need to thrive. In the next chapter, we will explore what the future of inclusive interfaces looks like and what steps can be taken to ensure that they are accessible to everyone.

Chapter Six

The Future of Inclusive Interfaces: What Lies Ahead and Within for Vulnerable Populations?

As technology continues to advance, it is crucial that we consider the future of inclusive interfaces and what it means for vulnerable populations. The design of accessible and inclusive interfaces has come a long way, but there is still much work to be done. With the increasing use of artificial intelligence and machine learning, the potential for creating interfaces that cater to the unique needs of disabled individuals is growing. However, this also means that we need to be vigilant about preventing the perpetuation of harmful biases and exclusionary practices. In this chapter, we will explore the possibilities and potential pitfalls of the future of inclusive interfaces, and what it means for vulnerable populations.

A look into the future of inclusive interfaces for vulnerable populations

As technology continues to advance at an unprecedented pace, the future of inclusive interfaces for vulnerable populations is becoming increasingly exciting. There are already numerous innovations and developments on the horizon that have the potential to make a significant impact on the lives of people with disabilities.

One area of development is the rise of wearable technology, which has already begun to make a significant impact in the lives of people with disabilities. Wearable technology like smartwatches and fitness trackers can be used to monitor vital signs and track fitness progress, helping individuals with disabilities to manage their health more effectively.

Smartwatches and fitness trackers can provide notifications regarding vital signs such as heart rate, blood pressure, and oxygen saturation, which is particularly helpful for individuals with conditions like heart disease or asthma. Additionally, these devices can track exercise and help users set and achieve health goals, such as increasing daily steps or monitoring calorie intake. Without access to wearable technology, vulnerable populations may be at a disadvantage in terms of managing their health and wellness, and may not have the same level of support and resources as those who can afford these devices. It would be unfair to deny them access to these technologies, as it would be like giving two devices to those who don't need them and none to those who do. Therefore, promoting the use of wearable technology for vulnerable populations is an important step towards ensuring equal access to healthcare resources and improving overall health outcomes.

Another area of development is the rise of artificial intelligence and machine learning, which has the potential to revolutionize the way that people with disabilities interact with technology. As AI becomes more advanced, it will be able to recognize and respond to a wider range of user needs, making technology more accessible for everyone.

If everyone who is able to access multiple artificial intelligence assistants and machine learning tools while the handicapped have no access to either, it would create a significant divide and disadvantage for the disabled community. This would not only be unfair, but it would also perpetuate existing inequalities and discrimination against disabled individuals.

Access to technology can greatly enhance the quality of life for people with disabilities. For example, voice-activated assistants and other assistive technologies can help individuals with physical disabilities to perform daily tasks more easily and independently. Without access to such technologies, individuals with disabilities would be more dependent on others and would have limited opportunities for self-sufficiency and autonomy.

Moreover, the lack of access to artificial intelligence assistants and machine learning tools would result in a lack of equal opportunities for education, employment, and social participation for the disabled community. This would create a further divide between the able-bodied and disabled populations and would hinder progress towards a more inclusive and equitable society.

Additionally, the increasing popularity of virtual and augmented reality technology is providing new opportunities for people with disabilities to experience the world in new ways. For example, VR technology can be used to simulate environments that are difficult or impossible to access for people with disabilities, while AR technology can be used to provide real-time information and assistance in a variety of contexts.

Once, there were two children named Jack and Alex. Jack was a healthy and able-bodied child, while Alex was a special needs child who used a wheelchair and had limited mobility. Jack had access to the latest virtual and augmented reality technology and was able to explore different environments, both real and simulated, with ease. He could experience the thrill of skydiving, swim with dolphins, and even visit far-off places like the Amazon rainforest.

On the other hand, Alex had no access to this technology and often felt left out. He would watch in envy as Jack would come to school and show off his latest VR headset or AR glasses, knowing that he would never have the same opportunities. Alex's limited mobility meant that he couldn't easily visit new places or experience new things like Jack could, and this made him feel isolated and left out.

One day, Jack invited Alex to try out his latest VR headset, hoping that it would help his friend experience new things like he did. Alex put on the headset and suddenly found himself in a completely different world. He was able to fly over the Grand Canyon, visit the Great Wall of China, and explore the depths of the ocean, all from the comfort of his wheelchair.

For the first time, Alex felt like he was no longer limited by his disability. The VR technology allowed him to experience things that he never thought possible, and it helped him feel like he was part of a larger world. The technology even had special features that helped him navigate through the virtual environment more easily, such as voice commands and eye tracking.

From that day on, Jack and Alex would use the VR technology together, exploring new worlds and experiencing new adventures. Jack was happy to share his technology with his friend, and Alex felt grateful for the opportunity to experience the world in a whole new way. Even though they had different abilities, they were able to enjoy the same technology and have fun together.

Overall, the future of inclusive interfaces for vulnerable populations is bright, with numerous advancements on the horizon that have the potential to make a significant impact on the lives of people with disabilities. As technology continues to advance and become more accessible, it is likely that we will see even more innovations and developments in the years to come.

The potential benefits and drawbacks of emerging technologies

Emerging technologies have the potential to significantly impact the lives of people with disabilities, but they also come with potential benefits and drawbacks. One potential benefit is the development of advanced prosthetics that are more intuitive, responsive, and capable of integrating with the user's body. These advancements have already allowed people with disabilities to perform tasks that were once impossible, such as running, dancing, and even playing musical instruments.

Another emerging technology with potential benefits is brain-computer interfaces, which allow individuals to control machines with their thoughts. This technology has the potential to significantly improve the quality of life for people with disabilities by providing greater independence and control over their environment.

However, there are also potential drawbacks to emerging technologies. One concern is the potential for technologies to perpetuate or even exacerbate existing inequalities. For example, if emerging technologies are only accessible to those who can afford them or have the technical knowledge to use them, they may widen the gap between the haves and have-nots, leaving those with disabilities further behind.

Another potential drawback is the risk of over-reliance on technology. While technology can be a powerful tool, it can also lead to complacency and reduced self-reliance if individuals become too reliant on it. For example, if a person with a visual impairment relies too heavily on a navigation app, they may not develop the skills necessary to navigate unfamiliar environments without it.

In addition, emerging technologies may also bring new ethical concerns, such as issues of privacy, consent, and control. As technology advances, it is essential to ensure that individuals with disabilities have the same rights and protections as everyone else.

Ultimately, the benefits and drawbacks of emerging technologies depend on how they are developed, implemented, and regulated. It is essential to approach these technologies with a critical eye and ensure that they are designed in a way that promotes inclusivity and equality for all, rather than perpetuating existing inequalities.

The importance of collaboration and community engagement

in creating safe and proven inclusive digital spaces

As we continue to develop and implement emerging technologies in the realm of inclusive interfaces, it is crucial to emphasize the importance of collaboration and community engagement. In order to create truly safe and proven inclusive digital spaces, we must work together and engage with the communities that these technologies will impact.

Collaboration is key in developing effective solutions for vulnerable populations. Bringing together diverse groups of individuals, including people with disabilities, technology developers, healthcare professionals, and policymakers, can lead to a more comprehensive understanding of the needs and challenges faced by those with disabilities. By working collaboratively, we can develop more effective and inclusive solutions that are better suited to the needs of those they are meant to serve.

Community engagement is also essential in creating safe and proven inclusive digital spaces. It is important to involve the communities that will be impacted by these technologies in the design and development process. By engaging with these communities, we can gain a better understanding of their unique needs and challenges, and work to create solutions that are more responsive and effective.

One example of collaboration and community engagement in creating safe and proven inclusive digital spaces is the development of the Web Content Accessibility Guidelines (WCAG) by the World Wide Web Consortium (W3C). The WCAG provides a set of technical standards for web content accessibility, such as ensuring that content can be navigated using a keyboard instead of a mouse, providing descriptive text for images, and using clear and simple language for written content.

The development of the WCAG involved a collaborative effort between individuals with disabilities, accessibility experts, and industry representatives, and included public feedback and comments to ensure that the guidelines were comprehensive and effective. This collaborative approach helped to ensure that the WCAG was not only technically sound but also responsive to the needs of disabled users.

Today, the WCAG is widely recognized as the gold standard for web accessibility, and many organizations and governments around the world use it as a benchmark for ensuring that their digital content is inclusive and accessible to all users, regardless of ability.

There are a number of different strategies that can be used to engage with communities and ensure that their needs are taken into account in the development of inclusive interfaces. One effective approach is to establish partnerships with community organizations that serve people with disabilities. These organizations can provide valuable insights and feedback on the design and development of new technologies, as well as help to connect developers with members of the community who can provide firsthand perspectives on their needs and experiences.

Another important strategy is to involve people with disabilities in the design and development process as much as possible. This can be done through a variety of means, including focus groups, user testing, and co-design processes. By involving people with disabilities in the development of inclusive interfaces, we can ensure that their needs are being addressed and that the solutions we create are more effective and responsive to their unique challenges.

Ultimately, the success of inclusive interfaces depends on our ability to work collaboratively and engage with the communities that they will impact. By prioritizing collaboration and community engagement, we can develop more effective and inclusive solutions that are better suited to the needs of vulnerable populations.

Reflections on the impact of "Sensory Enhancements for a More Connected World" on the conversation around inclusive interfaces for vulnerable individuals

The concept of "Sensory Enhancements for a More Connected World" has the potential to transform the conversation around inclusive interfaces for vulnerable individuals. By prioritizing the needs and experiences of disabled people in the design and development of digital interfaces, we can create technology that is not only accessible but also empowering for this population. This approach can lead to a more diverse and inclusive digital landscape that better serves the needs of all users.

One of the most significant impacts of "Sensory Enhancements for a More Connected World" is that it shifts the focus from accessibility as an afterthought to accessibility as a core aspect of the design process. This approach means that disabled individuals are no longer viewed as a niche market or an afterthought, but as an integral part of the user base. This shift in thinking can help to reduce the stigmatization and marginalization of disabled individuals, and create a more inclusive and welcoming digital environment.

Furthermore, "Sensory Enhancements for a More Connected World" can encourage collaboration and partnership between disabled individuals and technology developers. By engaging with disabled communities and working to understand their needs and experiences, technology companies can create interfaces that are not only accessible but also useful and enjoyable for this population. This approach can help to break down the barriers that often exist between technology developers and disabled communities, and create a more collaborative and inclusive design process.

Overall, the impact of "Sensory Enhancements for a More Connected World" on the conversation around inclusive interfaces for vulnerable individuals is significant. By prioritizing accessibility and inclusivity in the design and development of digital interfaces, we can create technology that better serves the needs of all users, regardless of ability. This approach can help to create a more diverse, equitable, and inclusive digital landscape, and ensure that everyone has access to the benefits and opportunities of technology.

Chapter Seven

Enriching Open Data Circumstances for Vulnerable Individuals

Have you ever wondered why some butterflies have wings that look like leaves? Why do animals in the zoo have enrichment tools in their cages like boxes and toys? Why do people who can't walk have devices within their reach? Why are grocery stores open, yet some of us are still unable to afford basic necessities? These questions all point to the need for enrichments in the lives of vulnerable individuals. In this chapter, we explore the concept of enriching open data circumstances for vulnerable individuals to reduce the risk of Sensory Enhancements for a More Connected World. Enrichment involves co-creating circumstances that provide a stimulating life for our vulnerable population, with opportunities to engage in natural dreams, be more active, and have more control over their internal environment. Enrichment helps satisfy both the physical, social, and psychological needs of disabled people and allows them to make more and healthier choices. Plus, it helps the outside world help itself.

The physical adaptations that can be made to technology, devices, and assistive tools are crucial for enriching the lives of vulnerable individuals. These adaptations can allow individuals with disabilities to have greater independence, control, and participation in daily activities.

The Importance of Physical Adaptations

One of the most significant advances in physical adaptations has been the development of smart homes. Smart homes are designed to be accessible and controllable by individuals with disabilities. They can be equipped with voice-activated assistants, sensors, and remote controls that allow users to control lighting, temperature, appliances, and security systems. For example, individuals with mobility impairments can use smart home technology to control their home environment without needing to move around physically.

Once there was a woman named Maria, who had a spinal cord injury and used a wheelchair. She lived alone in a small apartment and struggled with maintaining a healthy lifestyle due to her physical limitations. Maria often found it challenging to keep up with regular exercise routines and struggled to prepare healthy meals for herself.

However, things changed for her when she moved to a smart home designed specifically for people with disabilities. The home was equipped with various devices that helped her manage her daily routine and improve her health. She had a smart kitchen that had modified appliances, which allowed her to cook and prepare nutritious meals with ease.

The smart home also had an exercise room with adapted equipment, which she used to workout regularly. Additionally, Maria had wearable technology that monitored her vitals, reminding her to take breaks when needed and track her progress.

The smart home also had an integrated system that reminded her to attend regular checkups with her dentist and optometrist, which helped her maintain good oral and visual health. Maria was able to live more independently, and the smart home also provided her with a sense of safety and security.

As a result of these physical adaptations, Maria was able to take control of her health and live a more fulfilling life. She felt more confident and empowered to make healthy choices and take care of herself.

Data Ability Care is crucial for ensuring that vulnerable individuals are able to fully participate in digital spaces and access the information they need to thrive. This section focuses on the importance of modifying data to better serve the needs of people with disabilities.

One of the main considerations in Data Ability Care is data privacy and security. Vulnerable individuals often have sensitive personal information that needs to be protected, and they may face higher risks of data breaches or exploitation. Therefore, it is essential to implement ethical data practices that protect these populations from harm. This may include encryption, secure storage, and access controls to ensure that only authorized personnel can access sensitive data.

Another important aspect of Data Ability Care is ensuring that data is accessible to people with disabilities. This may involve adapting data formats, using accessible software, and providing alternative formats such as audio or braille. Additionally, open data initiatives can be leveraged to create more inclusive digital spaces. Open data allows for the sharing of information and resources, which can help to bridge the digital divide and ensure that all individuals have equal access to information.

One example of the benefits of Data Ability Care is the use of assistive technologies that can improve accessibility and inclusion. For instance, screen readers, text-to-speech software, and captioning services can all help individuals with visual or auditory impairments to access information online. Other tools like alternative input devices, voice recognition software, and eye-tracking technology can help individuals with physical disabilities to navigate digital interfaces.

Another example of Data Ability Care that benefited everyone, especially the vulnerable, is the implementation of closed captioning on television shows and movies. Closed captioning involves displaying written text on the screen to provide a transcript of the audio for people who are deaf or hard of hearing. This modification of data not only benefits those with hearing impairments, but also people who have difficulty understanding spoken language, such as those who are learning a new language or have cognitive impairments.

By implementing closed captioning, television shows and movies become more accessible to a wider audience, increasing inclusivity and allowing vulnerable individuals to fully participate in popular culture. This not only enriches their lives but also creates a more diverse and accepting society as a whole. Additionally, closed captioning has the potential to improve literacy skills in children who may struggle with reading or have learning disabilities.

Furthermore, the implementation of closed captioning also has the potential to improve data privacy and security for vulnerable individuals. By using ethical data practices and ensuring that closed captioning is accessible to all without compromising their privacy, we can create a more secure and inclusive digital environment for everyone.

Ultimately, Data Ability Care is about ensuring that data is accessible and inclusive for everyone, regardless of ability or disability. By implementing ethical data practices, adapting data formats and technologies, and leveraging open data initiatives, we can create more inclusive digital spaces that benefit all members of society.

Creating Private Social Groups

Social isolation can have a significant impact on the mental and physical health of vulnerable individuals. Social exclusion can lead to a lack of self-esteem, depression, and anxiety. Therefore, it is important to create private social groupings that allow individuals to connect with others who share similar experiences.

Technology has made it easier to create and maintain these social groups. Online communities, social media platforms, and virtual events can all provide opportunities for individuals with disabilities to connect with others in a safe and private environment. By creating these social groups, vulnerable individuals can find support and build relationships that can help them cope with their challenges and achieve their goals.

One example of this is the creation of online support groups for individuals with chronic illnesses. These groups provide a platform for people to connect with others who are going through similar experiences, share their stories, and offer support and advice. The anonymity and privacy provided by online groups make it easier for individuals to open up about their struggles and seek help without fear of judgment or stigma.

Moreover, virtual events and activities can also be organized for these social groups. For instance, an online cooking class can be arranged where people can learn to cook nutritious meals together while socializing. Other activities like book clubs, movie nights, or game nights can also be organized to provide entertainment and foster social connections.

In conclusion, creating private social groupings is crucial for the well-being of vulnerable individuals. By using technology to connect people and provide opportunities for socialization, we can help mitigate the negative effects of social isolation and improve mental health and overall quality of life.

Super Increasing Sensory Detections

Super Increasing Sensory Detections is an important area

to consider when it comes to enriching the lives of vulnerable individuals. Many people with disabilities face challenges when it comes to sensory processing and integration. Some individuals may have difficulty processing sensory information, while others may have difficulty integrating sensory information from multiple sources. These challenges can lead to feelings of isolation, disconnection, and frustration.

Jenny is a 25-year-old woman with autism. She has always struggled with sensory processing, and it has caused her a lot of frustration and isolation throughout her life. She finds loud noises overwhelming, certain textures uncomfortable, and bright lights too intense. As a result, she has avoided crowded places and social events that involve a lot of sensory stimulation.

Jenny's family noticed her struggles and decided to look for ways to help her. They came across a local center that specializes in providing multisensory experiences for individuals with disabilities. The center had various rooms that were designed to stimulate different senses, such as a room filled with soft textures, a room with calming music, and a room with gentle vibrations.

At first, Jenny was hesitant to try the center, but with encouragement from her family, she decided to give it a chance. To her surprise, she loved it. She found the different sensory experiences calming and enjoyable. For the first time, she felt connected to the world around her in a way that she had never experienced before.

Jenny's family also looked into technology that could help her with sensory processing. They found noise-canceling headphones that she could wear when she went out in public. The headphones helped her block out overwhelming noises and feel more comfortable in crowded places.

Thanks to these adaptations and technology, Jenny now feels more connected to the world and is no longer limited by her sensory challenges. She can attend social events and feel comfortable in public places, which has opened up new opportunities for her.

Fortunately, advancements in technology have made it possible to provide sensory enhancements that can improve the lives of more people with disabilities. For example, audio and visual enhancements can be used to provide a more immersive experience for people with hearing or visual impairments. These enhancements can include captioning and audio descriptions for videos, or audio feedback for individuals who may not be able to see or hear certain sounds.

Carla is a college student who has a hearing impairment. She has always struggled to hear lectures in class, even with the help of hearing aids. However, her college recently started offering a new technology called real-time captioning, which displays live captions of the lecture on her laptop as the professor speaks. This has been a game changer for Carla, as she can now easily follow along with the lecture and participate in class discussions without feeling left out.

In addition to real-time captioning, Carla's college has also started offering visual enhancements during lectures. For example, instead of relying solely on written slides, professors are now using more images and videos in their presentations to help illustrate concepts. They also use more visual aids, such as diagrams and graphs, to help convey information in a way that is easier to understand.

Thanks to these audio and visual enhancements, Carla has been able to fully participate in her classes and feel more connected to her peers and professors. She is also able to learn more effectively and retain information better, which has helped her excel academically. Overall, these small but impactful changes have made a big difference in Carla's life and have helped her feel more included and supported in her academic pursuits.

Another way to provide multisensory experiences is through the use of haptic feedback. Haptic feedback involves the use of vibrations, pressure, or other tactile sensations to provide information to the user. This can be particularly useful for individuals who may have difficulty processing or integrating visual or auditory information. For example, a person with a visual impairment may use a haptic feedback device to navigate through an unfamiliar environment.

Carla is a 12-year-old girl who is visually impaired. She often struggles to navigate unfamiliar places and feels isolated from her sighted peers. However, her school recently installed a new technology called haptic feedback that allows her to experience multisensory feedback in real-time.

Carla was amazed by the new technology. She was able to navigate through the school hallways using haptic feedback, which provided her with tactile sensations that helped her understand her surroundings. For example, when she approached a corner, her device would vibrate, signaling that she needed to turn.

When she reached a stairwell, her device would provide her with pressure feedback, signaling that she needed to descend.

The haptic feedback technology not only helped Carla navigate her school, but it also allowed her to participate in activities that were previously inaccessible. For example, during science class, Carla was able to touch and feel objects like rocks and minerals with the help of haptic feedback. She was also able to participate in physical education classes that involved team sports, as her device provided her with feedback about the position and movement of other players.

Overall, haptic feedback technology helped Carla feel more connected to her environment and allowed her to participate in activities that she previously thought were impossible.

Additionally, providing sensory-rich experiences can also help people with disabilities feel more connected to the world around them. For example, multisensory art installations or immersive experiences can provide a sense of presence and engagement that may be difficult to achieve through traditional means.

Mark was a seasoned news reporter who had been in the industry for over a decade. He prided himself on his ability to dig up information, conduct interviews, and deliver the news with accuracy and speed. However, when it came to using new technology like ChatGPT, he had always been a bit hesitant.

He had tried using the language model in the past, but it often "torpedoed" him with foreign data, making it difficult to get the information he needed. Mark felt like the machine was working against him, and it frustrated him to no end.

As technology continued to advance, Mark knew he couldn't avoid ChatGPT forever. He needed to learn how to use it to keep up with the competition and deliver the news his viewers wanted. However, he didn't know where to start or how to make it work for him.

One day, Mark heard about Sensory Enhancements for a More Connected World, a new movement aimed at making data more accessible and inclusive for vulnerable populations. He was intrigued and decided to attend a workshop on the subject.

There, he learned about the importance of adapting technology to meet the needs of all individuals, including those with disabilities. He discovered how ChatGPT could be modified to filter out foreign data and deliver information in a way that was more manageable for him.

Mark also learned about the importance of data privacy and security for vulnerable populations and how ethical data practices could protect them. He realized that by being patient and taking the time to understand the technology, he could use it for work and his viewers without fear.

After the workshop, Mark started incorporating ChatGPT into his reporting, using it to gather information and conduct interviews. He found that with Sensory Enhancements for a More Connected World, the language model "quad-party" plugin made it much easier to use and provided him with the information he needed to deliver the news to his viewers.

Mark's reporting improved, and he became more confident in his ability to use technology to enhance his work. He was grateful for the Sensory Enhancements for a More Connected World movement, which not only helped him but also made data more accessible and inclusive for everyone.

Overall, Super Increasing Sensory Detections is a crucial area to consider when it comes to enriching the lives of vulnerable individuals. By providing sensory enhancements and multisensory experiences, we can help individuals with disabilities feel more connected, engaged, and fulfilled in their lives.

Summary of Sensory Enhancements for a More Connected World

This book focused on the concept of Sensory Enhancements for a More Connected World and explores how data can be enriched to better serve the needs of vulnerable individuals. It examined the importance of creating inclusive digital spaces, modifying technology and devices to meet specific needs, implementing ethical data practices, and providing opportunities for socialization and community building. The book also explored the benefits of sensory enhancements and multisensory experiences for people with disabilities.

Through a series of examples and case studies, the book demonstrated how Sensory Enhancements for a More Connected World can lead to better quality of life for vulnerable individuals and benefit society as a whole.Section 504 and Title II nondiscrimination standards were significant milestones in ensuring equal access and opportunities for individuals with disabilities. These standards paved the way for a more inclusive society and enabled people with disabilities to participate more fully in everyday life. In the same vein, this Sensory Enhancements for a More Connected World book explores how technology and innovation can continue to break down barriers and create a more accessible and equitable world for all.

In conclusion, Sensory Enhancements for a More Connected World is not only about creating accessible and inclusive digital spaces but also about promoting social inclusion, providing opportunities for physical, mental, and emotional well-being, and enhancing the overall quality of life for vulnerable individuals. By enriching data in a thoughtful and responsible manner, we can ensure that everyone has the opportunity to participate fully in society and live their lives to the fullest.

Afterword

Acknowledgments

I would like to express my deepest gratitude to all of the individuals who helped make this book possible. To my family and friends, thank you for your unwavering support and encouragement throughout this process. To the experts and advocates who generously shared their knowledge and experiences with me, thank you for your time and insights.

I would also like to thank the good Lord Jesus, the team at united CLU adaptive paranting, and the team at MajorLeagueFamily.com for their collaborations and expertise in bringing this book to fruition. Finally, I want to express my gratitude to the many individuals with disabilities who have taught me so much about resilience, determination, and the power of community. Your voices and stories have inspired me to continue working towards a more just and equitable world.

About Author

Meet Least Commanded (Matthew 5:19) Author Aaron W. Wemple, the self-proclaimed detoxifier of AI data metrics and developer of nuclear family severance valueation methods, including the hybrid family law's constructive child safety files. With a heart bigger than his brain, Aaron is committed to earning problems away for families and friends in crisis. And if you're in a high-conflict situation, Aaron is your guy! He'll create a Child Safety File faster than you can say "MajorLeagueFamily.com".

But don't let his dedication
to justice fool you -
Aaron's also a talented writer, community organizer, and visionary entrepreneur. His innovative approach to business is like nothing you've ever seen before - unless, of course, you've seen a unicorn riding a rainbow.

And the accolades keep rolling in! News leaders, book reviewers, and government organizations have all recognized Aaron's next-generation data metrics approach. We're not sure what that means, but it sounds pretty impressive.

Join Aaron the team's quest to save the world through AI for good. He's already hashtagging his way to success with #LeastCommanded, #UpliftingFamilies, #ProtectingVulnerableIndividuals, #AIForGood, and #PracticeSmall. Who needs actual results when you have catchy hashtags?

Also By

Change the Game: One Simple New Deal Can Help Any Family Change Any Industry (Series)

A Light in the Dark: Ace and Crush's Story of Courage and Determination

Adaptive CLU Parenting Guide

Adorning Your Permanent Losses So the Future You Adores It (CLU Self-Help Guide)

Breaking the Mold: Transforming Legal Learning with Professional Law Caveats - (Family Law 2.0 Second Edition)

Co-open AI: Redefining the Future of AI-UI Interfaces

Dealtown Tailoring: Surviving Anguish or Success in America?

Faithfinder: Inclusive Holiday Stories Empower Readers to Find Their Superfaith!

Family Law 2.0: Hatching a Superslate (Family Law 2.0 First Edition)

Future 12: A Priceless Guide to Creating Fair and Unbiased A.I. Systems

Impossible Data: How Making Almighty Data Cavityism History Can Help Us Live A Better Life

IR Physics 2

Parenting 2.0 For Un-Haters

Revitalizing Education: The Role of Academic Law Caveats in Student Growth

Revitalizing Inclusive Socioeconomics: The Role of SocioEcTech Justice on Human Growth

Sensory Enhancements for a More Connected World

The Psychology Doctor: Understanding the Intersection of Organic Childhood Psychology and Science

Thriving Against the Odds: How to Innovate in Deprived Communities

Outside The Series:

American Pros

Bridgd Forum @
cleanlawunion.com

Family Law 2.0

I R "Freestyle" Physics – The Study of Transformation and/or Transfiguration

Law & Side-Law: Family Hostility / Family Empathy

Quantum Layer Taylors Blog @
majorleaguefamily.com

The Law Doctor

Upbright USA

Made in the USA
Monee, IL
24 November 2024

71052973R00131